数字故事

THE NUMBER STORY

SMALL BOOK ONE

ENGLISH - SIMPLIFIED CHINESE

*Numbers Teach Children
Their Number Names*

written and illustrated by

MISS ANNA

Early Reader Edition of *The Number Story 1*
Bronze Medal Winner, 2016 Wishing Shelf Book Award

LUMPY PUBLISHING

Cover by | Lumpy Publishing
Layout by | Lumpy Publishing
Translated by Carol Gwo
Coloring by Jieeun Woo and Maria

Library of Congress Control Number: 2018902040

Names: Miss Anna, author.
Title: Number story : numbers teach children their number names / Miss Anna.
Description: Portland, OR: Lumpy Publishing, 2018.
Identifiers: ISBN 978-1-949320-18-3 | LCCN 2018902040
Summary: The pictures and rhymes present stories which introduce numbers 0-10.
Subjects: LCSH Numeration--English--Pictorial works--Juvenile literature. | BISAC JUVENILE NONFICTION /
Languages: English—Simplified Chinese
Classification: LCC QA141.3 .M57 2018 | DDC 513—dc23

Publisher: Lumpy Publishing
Website: www.missannabooks.com
Email: missanna@missannabooks.com

Paperback: ISBN 978-1-949320-18-3
Printed in the U.S.A.   1 3 5 7 9 10 8 6 4 2

Want to learn our number names?

想学习数字名称吗？

It is very easy and a lot of fun!

这很容易且有趣！

Say-along our little jingle

跟我们唱儿歌。

starting from Number One!

我们从一开始!

1
ONE looks like my one finger.
一 就像一根手指。

ONE!

2

TWO   trails a tail.

二　像一条小尾巴。

嗖 嗖~
SWISH!
SWISH!

3

THREE has bumps.

三　有不平的坑洼。

BUMPY! 顛簸!

4

FOUR carries a sail.

四 杨帆起航。

4
A SAIL!
起航！

5
FIVE   is a racing track.
五 像是一条跑道。

VROOM
哦也！

6

SIX   curves like a snail.

六 又曲又弯像蜗牛。

蜗牛！

A SNAIL!

7

SEVEN has a sharp angle.

七 有尖锐的角。

OUCH!
哎哟!

8

EIGHT   is rollercoaster rails.

八 相似过山车轨道。

呦呼！
YIPPEE!

9

NINE is a bubble on a stick.

九 泡泡在棍子的顶部。

A BUBBLE! 泡泡!

TEN is an eye of a whale.

十　鲸鱼的一只眼睛。

眨眼！
WINK!

And
和

0

ZERO   is an empty pail.

零 像是一个空桶。

IT'S
EMPTY!
空的！

Thank you for playing with us today.

We had a lot of fun too!

感谢你今天和我们一起玩。

我们也有很多乐趣！

We are your Number friends,
Zero to Ten,
Who will be here for you~
我们是你的数字朋友，从零到十。
我们在这里等待你的光临。

Bye-bye now!
See you again soon!
再见！
期待很快在见到你。

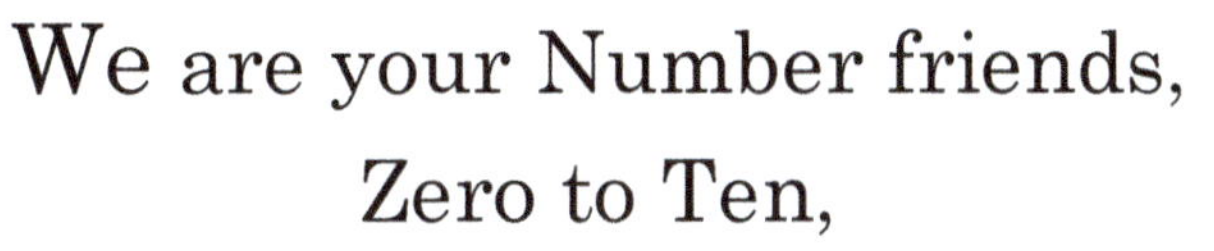

The Numbers are *SINGING* too!

To sing-a-long, look for Miss Anna Number Story
at your favorite music store like iTUNES.

MP3

Numbers 0-10
IDENTIFYING & COUNTING

Numbers 11-20
& Ordinals
first, second, third...

Numbers 0-100
& Place Values
ones, tens, hundreds...

About Clocks
& Telling Time
hours, minutes, seconds

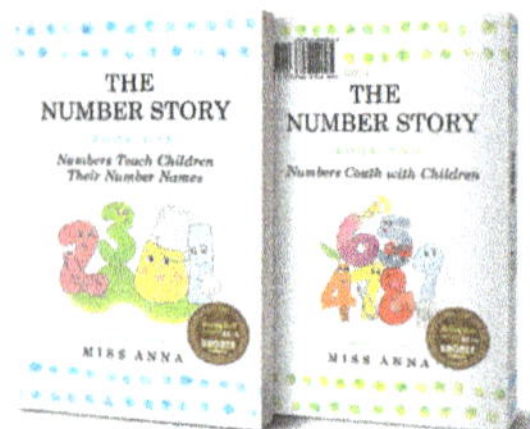

Number Story 1 & 2
isbn: 978-0-996216-48-7

Number Story 3 & 4
isbn: 978-1-945977-01-5

Number Story 5 & 6
isbn: 978-1-945977-06-0

Number Story 7 & 8
isbn: 978-1-949320-40-4

For more Miss Anna books to love,
visit us at

www.missannabooks.com

Numbers are working hard all over the world!
Come Travel the World with Us!